Essential Trigonometry

A Self-Teaching Guide

Tim Hill

Questing Vole Press

Essential Trigonometry: A Self-Teaching Guide
by Tim Hill

Editor: Kevin Debenjak
Proofreader: Diane Yee
Compositor: Kim Frees
Cover: Questing Vole Press

Second Edition

Contents

1   A Few Basics

Before jumping into trigonometry, you need to know a few basic facts from algebra and geometry. Feel free to skip or skim this chapter if you're already familiar with the material.

Angles in a Triangle Sum to 180 Degrees

The **degree** is the customary unit of measure for angles. A **right angle** measures 90 degrees (90°), a **straight angle** (semicircle) measures 180°, and 360° sweeps a complete circle. Figure 1.1 shows some angles, measured in degrees.

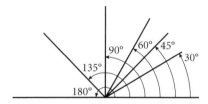

Figure 1.1 Angles measured by counterclockwise rotation from the positive x-axis

In Figure 1.2, a **transversal** crosses a pair of parallel lines. The corresponding angles are equal and the alternate interior angles are equal.

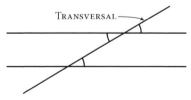

Figure 1.2 A transversal cutting two parallel lines

The sum of the angles in every triangle equals 180° (Figure 1.3).

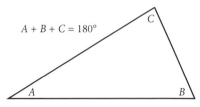

$A + B + C = 180°$

Figure 1.3 The angles of a triangle sum to a straight angle (180°)

You can prove this fact by using the preceding results, coupled with an auxiliary parallel line (Figure 1.4).

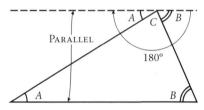

PARALLEL

180°

Figure 1.4 The dotted auxiliary line forms angles congruent to A and B

Consequently, in every triangle an exterior angle equals the sum of the opposite interior angles (Figure 1.5).

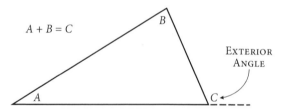

$A + B = C$

EXTERIOR
ANGLE

Figure 1.5 An exterior angle is always larger than either opposite interior angle

In a right triangle the sum of the acute angles equals 90° (Figure 1.6).

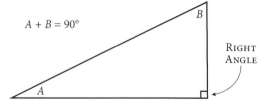

$A + B = 90°$

RIGHT
ANGLE

Figure 1.6 The legs of a right triangle are the two sides that meet at 90°

Similar Triangles Have Proportional Sides

Two triangles are **similar** if one is a zoomed version of the other—that is, they have the same shape but different sizes (Figure 1.7).

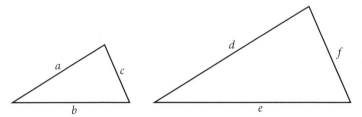

Figure 1.7 Similar triangles

Formally, triangles are similar if their corresponding angles are equal. Because the angles of every triangle sum to 180° (page 1), two triangles are similar if two pairs of their corresponding angles are equal, which implies that the third pair are equal too.

Equivalently, the corresponding sides of similar triangles have lengths in the same ratio:

$$\frac{a}{d} = \frac{b}{e} = \frac{c}{f}$$

You can rearrange the ratios by using elementary algebra. The equation $\frac{a}{d} = \frac{b}{e}$, for example, can be rewritten as $\frac{a}{b} = \frac{d}{e}$.

The Pythagorean Theorem

In a **right triangle** (or **right-angled triangle**), one angle is a right angle, a 90° angle. The **Pythagorean theorem** states that in every right triangle the square of the **hypotenuse** (the longest side) equals the sum of the squares of the legs (Figure 1.8).

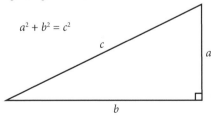

$$a^2 + b^2 = c^2$$

Figure 1.8 The Pythagorean theorem: $a^2 + b^2 = c^2$

The Pythagorean theorem has hundreds of proofs. The following algebraic proof uses a square with side c arranged inside four copies of a right triangle with sides a, b, and c. The side of the outer square, formed arranging the four replica triangles symmetrically, is $a + b$ (Figure 1.9).

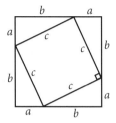

Figure 1.9 Algebraic proof of the Pythagorean theorem

The area of the outer (large) square equals the area of the four triangles plus the area of the inner (small) square:

$$(a+b)^2 = 4\left(\tfrac{1}{2}ab\right) + c^2$$

Expand and solve to get

$$a^2 + 2ab + b^2 = 2ab + c^2$$
$$a^2 + b^2 = c^2$$

Two Special Right Triangles

Two right triangles hold a special place in trigonometry, namely the 30°–60°–90° triangle and the 45°–45°–90° triangle (Figure 1.10). The latter is also called an isosceles right triangle.

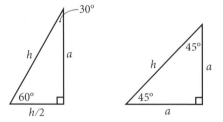

Figure 1.10 The 30°–60°–90° triangle and the 45°–45°–90° triangle

In a 30°–60°–90° triangle, we know from geometry that the side opposite the 30° angle is half the hypotenuse h. To find the length of leg a in terms of h, use the Pythagorean theorem:

$$a^2 + \tfrac{1}{4}h^2 = h^2$$
$$a^2 = \tfrac{3}{4}h^2$$
$$a = \tfrac{\sqrt{3}}{2}h$$

For a 45°–45°–90° triangle, use the Pythagorean theorem to solve for a in terms of h:

$$a^2 + a^2 = h^2$$
$$2a^2 = h^2$$
$$a = \tfrac{1}{\sqrt{2}}h = \tfrac{1}{2}\sqrt{2}h$$

The last step above rationalizes the denominator $\tfrac{1}{\sqrt{2}} = \tfrac{1}{\sqrt{2}} \cdot \tfrac{\sqrt{2}}{\sqrt{2}} = \tfrac{1}{2}\sqrt{2}$.

The Distance Between Two Points

If (x_1, y_1) and (x_2, y_2) are two points in the xy-plane, then the lengths of the legs of the right triangle in Figure 1.11 are $|x_1 - x_2|$ and $|y_1 - y_2|$, where $|a|$ denotes the **absolute value** of a (the non-negative value of a without regard to its sign; $|3| = 3$, $|0| = 0$, and $|-2| = 2$, for example).

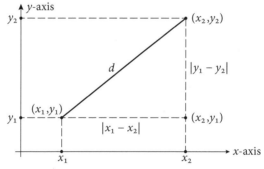

Figure 1.11 Derivation of the distance formula

Applying the Pythagorean theorem

$$d^2 = \left|x_1 - x_2\right|^2 + \left|y_1 - y_2\right|^2$$
$$= \left(x_1 - x_2\right)^2 + \left(y_1 - y_2\right)^2$$

yields the **distance formula**

$$d = \sqrt{\left(x_1 - x_2\right)^2 + \left(y_1 - y_2\right)^2}$$

The Unit Circle

The **unit circle** is a circle with a radius of one. In trigonometry, the unit circle is the set of all points (x, y) in the xy-plane whose distance from the origin $(0, 0)$ is equal to 1 (Figure 1.12).

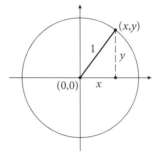

Figure 1.12 The unit circle

In Figure 1.12, $|x|$ and $|y|$ are the lengths of the legs of a right triangle whose hypotenuse has length 1. By the Pythagorean theorem, the equation of the unit circle is

$$x^2 + y^2 = 1$$

Functions

A **function** associates every number from a first set, called the **domain**, with another number in a second set, called the **range**, such that each element in the domain corresponds to *exactly one* element in the range. (Functions can be defined more generally to deal with other objects, but I consider only real numbers in this book.)

Functions are typically denoted by the letters f, g, and h. If f is a function and x is a number in the domain of f, then the number that f associates with x is denoted by $f(x)$ and is called the value of f at x. The symbol $f(x)$ is read "f of x."

If a function f is defined by the formula

$$f(x) = x^2$$

for every real number x, then the domain of f is the set of real numbers, and f is a function that associates every real number with its square.

To evaluate f at any number, square that number. For example,

$$f(3) = 3^2 = 9$$

$$f(-\tfrac{1}{2}) = (-\tfrac{1}{2})^2 = \tfrac{1}{4}$$

$$f(1 + a) = (1 + a)^2 = 1 + 2a + a^2$$

It's useful to think of a function as a machine or "black box" that when given an input x produces an output $f(x)$ (Figure 1.13). The same input must always produce the same output. Although each input has a unique output, a given output may result from more than one input. The inputs 3 and −3, for example, both produce the output 9 in the preceding example.

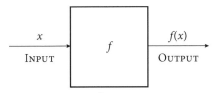

Figure 1.13 Function as machine

You can also think of a function as a mapping of one set of points (the domain) onto another set of points (the range) (Figure 1.14).

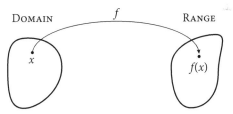

Figure 1.14 Function as mapping

When working with functions in the xy-plane, y is called a function of x, symbolized by

$$y = f(x)$$

Using the preceding example, $y = f(x)$ where $f(x) = x^2$. The letter f represents the rule or operation (squaring, in this case) that yields y when applied to x.

Graphs of Functions

The best way to visualize a function is by its **graph** (Figure 1.15).

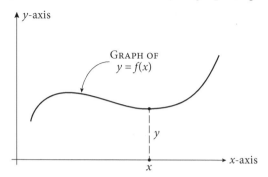

Figure 1.15 The graph of a function

When a function is defined by an equation in x and y, the graph of the function $y = f(x)$ is the set of points (x, y) in the xy-plane that satisfies the equation. The variable x is called the **independent variable** because it's free to take on any value in the domain, and y is called the **dependent variable** because its value depends on x.

Many types of functions exist, the most familiar being those defined by simple algebraic equations (Figure 1.16). If you're not using a computer, the best way to graph a function is to plot a few "interesting" points and then sketchily connect them according to the characteristics of the equation (linear, power, polynomial, logarithmic, and so on).

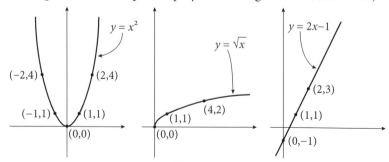

Figure 1.16 Graphs of $y = x^2$, $y = \sqrt{x}$, and $y = 2x - 1$

Problems

1. In a right triangle, the ratio of the measures of the acute angles is 4:1. What are these angles measured in degrees?

2. What is the area of an isosceles right triangle whose hypotenuse has length h?

3. The equal sides of an isosceles triangle are 2. If x is the base, express the area as a function of x.

4. A rectangle whose base has length x is inscribed in a fixed circle of radius a. Express the area of the rectangle as a function of x.

5. Find the distance between the points $(-7, 3)$ and $(1, -2)$.

6. Find the distance between the points (a, b) and (b, a).

7. If $f(x) = 5x^2 - 3$, find:
   (a) $f(-3)$ (b) $f(2)$ (c) $f(0)$
   (d) $f(-\sqrt{7})$ (e) $f(a + 3)$ (f) $f(5t)$

8. Compute and simplify the quantity $\dfrac{f(x+h)-f(x)}{h}$ for:

   (a) $f(x) = 5x - 3$
   (b) $f(x) = x^2$
   (c) $f(x) = 1/x$

9. Are any of the following pairs of functions equal?

   (a) $f(x) = x/x$, $g(x) = 1$
   (b) $f(x) = x^2 - 1$, $g(x) = (x + 1)(x - 1)$
   (c) $f(x) = x$, $g(x) = \sqrt{x^2}$
   (d) $f(x) = x$, $g(x) = (\sqrt{x})^2$

2 Radian Measure

In basic mathematics and everyday life, angles are measured in degrees (page 1), an arbitrary measurement inherited from ancient astronomers. In calculus and in most of science and engineering, the natural unit for measuring angles is the **radian**, which is defined in terms of how much arc an angle cuts off on a circle. Using radians leads to much tidier formulas than would be got by using degrees.

Consider an angle, denoted by the Greek letter θ (theta), placed at the center of a circle with radius r (Figure 2.1).

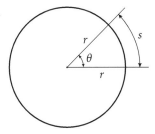

Figure 2.1 Arc subtended by the sides of the central angle θ

If s is the length of the arc **subtended** (cut off) by the sides of the angle θ, measured in the same linear units as the radius r, then the number of radians in θ is defined to be the value s/r, so

$$\theta = \frac{s}{r}$$

The length of a circular arc is therefore $s = r\theta$.

Note that an angle of 1 radian cuts off an arc that equals the length of the radius (Figure 2.2). In the unit circle (page 6), a central angle of 1 radian subtends an arc of length $s = r = 1$.

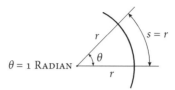

Figure 2.2 If $\theta = 1$ radian, then $s = r$

Converting Between Radians and Degrees

A 360° sweep draws a circle, and a 180° sweep draws a semicircle. The formula for the circumference c of a circle is

$$c = 2\pi r$$

where π is approximately 3.14159 and r is the radius of the circle. The length of a semicircular arc is πr (Figure 2.3).

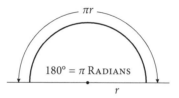

Figure 2.3 The length of a semicircular arc is πr or, equivalently, 180°

From the definition of a radian, we know that a semicircle represents $s/r = \pi r/r = \pi$ radians, giving the basic relation

$$\pi \text{ radians} = 180°$$

It follows that

$$1 \text{ radian} = \left(\frac{180}{\pi}\right)^{\circ} \approx 57.3° \quad \text{and} \quad 1° = \frac{\pi}{180} \approx 0.0175 \text{ radians}$$

The radians–degrees conversion formulas are

$$\theta \text{ radians} = \left(\frac{180\theta}{\pi}\right)^{\circ} \quad \text{and} \quad \theta° = \frac{\theta\pi}{180} \text{ radians}$$

By convention, the word "radian" is omitted in radian measure. The angle $\pi/6$ means an angle of $\pi/6$ radians, the angle 2 means an angle of 2 radians, and the angle θ means an angle of θ radians.

You may want to memorize the following conversions of commonly used angles.

Degrees	Radians
30°	$\pi/6$
45°	$\pi/4$
60°	$\pi/3$
90°	$\pi/2$
135°	$3\pi/4$
180°	π
270°	$3\pi/2$
360°	2π

To gain comfort with trigonometry, you should "think in radians" rather than mentally translate from degrees to radians (the same principle applies to learning a foreign language). Some key facts are:

- The angles of a triangle sum to π radians

- A right angle is $\pi/2$ radians

- Each angle of an equilateral triangle is $\pi/3$ radians

- The line $x = y$ in the xy-plane makes an angle of $\pi/4$ radians with the positive x-axis

- In a right triangle with a hypotenuse of length 1 and another side of length ½, the angle opposite the side with length ½ is $\pi/6$ radians

- One complete rotation around a circle is 2π radians

Exercise: A slice (like a slice of pizza) with angle θ radians inside a circle with radius r has area $\frac{1}{2}\theta r^2$. To see why radians yield cleaner formulas than degrees do, convert this formula to degrees.

Positive and Negative Angles

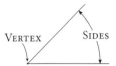

So far, we've defined an angle by its two **sides** (rays, or half lines) meeting at a common endpoint called a **vertex**. In trigonometry and calculus, however, angles are **directed**, meaning that one side is designated as the **initial side** and the other as the **terminal side**, and the angle is generated by rotating the initial side to the terminal side (Figure 2.4). An angle in **standard position** in the *xy*-plane has its vertex at the origin and its initial side lying along the positive *x*-axis.

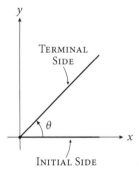

Figure 2.4 Generating an angle by rotation

Rotating the initial side **counterclockwise** to its terminal side generates a **positive angle**, whereas rotating **clockwise** generates a **negative angle** (Figure 2.5).

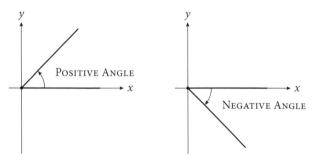

Figure 2.5 Rotation direction determines whether an angle is positive or negative

Example Angles

Figure 2.6 shows some positive angles in standard position.

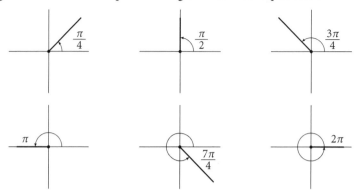

Figure 2.6 Positive angles in standard position

Figure 2.7 shows some negative angles in standard position.

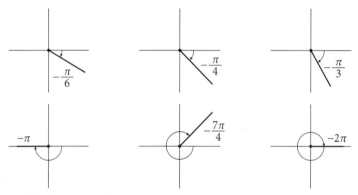

Figure 2.7 Negative angles in standard position

Angles greater than 2π can be generated by more than one rotation of the initial side (Figure 2.8).

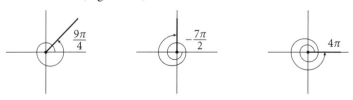

Figure 2.8 Angles greater than 2π

Coterminal Angles

Two angles with the same initial and terminal sides but possibly different rotations are called **coterminal angles**. Every angle has infinitely many coterminal angles because the rotation of the angle can be extended by one or more complete rotations of 2π, clockwise or counterclockwise. The resulting angle has the same initial and terminal sides as the original angle. Figure 2.9 shows a pair of coterminal angles.

Figure 2.9 Coterminal angles

Also note the coterminal angles in the figures in "Example Angles" earlier in this chapter:

- $\pi/4$ and $-7\pi/4$ and $9\pi/4$
- $\pi/2$ and $-7\pi/2$
- π and $-\pi$
- 2π and -2π and 4π

Problems

1. Convert the given angles from degrees to radians.

$$15° \qquad 150° \qquad 1500°$$

$$-36° \qquad -45° \qquad 7°$$

$$900° \qquad 1080° \qquad -110°$$

2. Convert the given angles from radians to degrees.

$$4\pi \qquad \frac{\pi}{9} \qquad 3$$

$$-\frac{2\pi}{3} \qquad \frac{3\pi}{2} \qquad 2\pi$$

$$\frac{7\pi}{6} \qquad \frac{14\pi}{6} \qquad -5\pi$$

3. For a 16-inch pizza, find the area of a slice with angle ¾ radians.

4. A particle travels counterclockwise on the unit circle from the point $(0, 1)$ to the endpoint of the radius that forms an angle of $5\pi/4$ radians with the positive x-axis. How far does the particle travel?

3   The Trig Functions

In elementary mathematics, the sine, cosine, and tangent of an acute angle θ are defined in terms of a right triangle with θ as one of its acute angles (Figure 3.1).

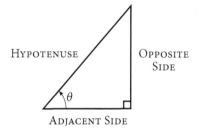

Hypotenuse   Opposite Side

θ

Adjacent Side

Figure 3.1 The right-triangle approach to trigonometry

The values of sin θ, cos θ, and tan θ are defined to be ratios of the sides of this triangle, where

$$\sin\theta = \frac{\text{opposite side}}{\text{hypotenuse}}$$

$$\cos\theta = \frac{\text{adjacent side}}{\text{hypotenuse}}$$

$$\tan\theta = \frac{\text{opposite side}}{\text{adjacent side}}$$

You can use the mnemonic SOH-CAH-TOA to remember these relationships: Sine = Opposite ÷ Hypotenuse, Cosine = Adjacent ÷ Hypotenuse, Tangent = Opposite ÷ Adjacent.

Defining Trig Functions on the Unit Circle

To define sin θ, cos θ, and tan θ sensibly, the right-triangle approach to trigonometry restricts θ to be positive and less than $\pi/2$ (90°). For most mathematical, scientific, and engineering purposes, however, θ must be unrestricted.

To remove this restriction, trig functions are defined on the unit circle (page 6). Figure 3.2 shows the unit circle $x^2 + y^2 = 1$ in the xy-plane. Generate θ—any number, large or small, positive or negative—by rotating the positive x-axis (initial side) counterclockwise by θ radians (if θ is positive) or clockwise by $-\theta$ radians (if θ is negative). The value of θ determines the endpoint (x, y) of the angle's terminal side.

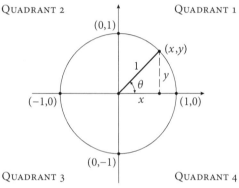

Figure 3.2 The unit-circle approach to trigonometry

The endpoint (x, y) can lie in any of four **quadrants**, with any combination of positive or negative x and y coordinates. Ignoring their algebraic signs, you can think of x and y as the base and the height of a right triangle whose hypotenuse is the terminal side of the angle θ. The six **trigonometric functions**, the three mentioned above along with the cotangent, secant, and cosecant, are defined to be

$$\sin\theta = y \qquad \tan\theta = \frac{y}{x} \qquad \sec\theta = \frac{1}{x}$$

$$\cos\theta = x \qquad \cot\theta = \frac{x}{y} \qquad \csc\theta = \frac{1}{y}$$

Note that the unit-circle definitions and the right-angle definitions of sin θ, cos θ, and tan θ are consistent when θ is a positive acute angle. The angle θ in Figure 3.2 is drawn to be equal to the angle θ in Figure 3.1, so that the two triangles are similar (page 3). In Figure 3.2, notice that

$$\sin\theta = y = \frac{y}{1} = \frac{\text{opposite side}}{\text{hypotenuse}}$$

$$\cos\theta = x = \frac{x}{1} = \frac{\text{adjacent side}}{\text{hypotenuse}}$$

$$\tan\theta = \frac{y}{x} = \frac{\text{opposite side}}{\text{adjacent side}}$$

The Essential Trigonometric Identities

The numbered identities, labeled 1–21 in the rest of this book, are the core relationships and properties of the trig functions. They're grouped logically to make them easier to find and remember. You don't actually *have* to memorize any of these identities—with a little algebra, they can all be derived by using the definitions and diagrams given earlier. Similar identities for the cotangent, secant, and cosecant also exist but they're mostly unimportant and omitted for brevity, in keeping with my goal of skipping the nonessentials.

Identities in Terms of Sine and Cosine

Sine and cosine are the two fundamental trig functions. If necessary, the other four trig functions can be expressed in terms of sine and cosine.

$$\tan\theta = \frac{\sin\theta}{\cos\theta} \tag{1}$$

$$\cot\theta = \frac{\cos\theta}{\sin\theta} \tag{2}$$

$$\sec\theta = \frac{1}{\cos\theta} \tag{3}$$

$$\csc\theta = \frac{1}{\sin\theta} \tag{4}$$

$$\cot\theta = \frac{1}{\tan\theta} \tag{5}$$

Identities for −θ

The angles θ and −θ are generated by the same rotation but in opposite directions, so that the endpoints of their terminal sides lie on the same vertical line (Figure 3.3).

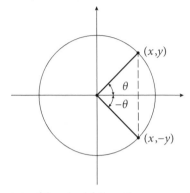

Figure 3.3 The endpoints of θ and −θ fall on the same vertical line

Replacing the angle θ by −θ yields

$$\sin(-\theta) = -\sin(\theta) \tag{6}$$

$$\cos(-\theta) = \cos(\theta) \tag{7}$$

$$\tan(-\theta) = -\tan(\theta) \tag{8}$$

(6) and (7) are apparent from Figure 3.3. (8) is derived by combining (1) with (6) and (7).

Pythagorean Identities

The equation of the unit circle is $x^2 + y^2 = 1$. If the terms are rearranged as $y^2 + x^2 = 1$, then this equation translates to

$$\sin^2 \theta + \cos^2 \theta = 1$$

The notations $\sin^2\theta$ and $\cos^2\theta$ mean $(\sin \theta)^2$ and $(\cos \theta)^2$. The related identities (all equivalent to each other) are

$$\sin^2 \theta + \cos^2 \theta = 1 \tag{9}$$

$$\tan^2 \theta + 1 = \sec^2 \theta \tag{10}$$

$$1 + \cot^2 \theta = \csc^2 \theta \tag{11}$$

(9) is equivalent to the Pythagorean theorem (page 3). To derive (10), divide (9) through by $\cos^2\theta$ to get

$$\left(\frac{\sin\theta}{\cos\theta}\right)^2 + \left(\frac{\cos\theta}{\cos\theta}\right)^2 = \left(\frac{1}{\cos\theta}\right)^2$$

To derive (11), divide (9) through by $\sin^2\theta$ to get

$$\left(\frac{\sin\theta}{\sin\theta}\right)^2 + \left(\frac{\cos\theta}{\sin\theta}\right)^2 = \left(\frac{1}{\sin\theta}\right)^2$$

Problems

1. Find cos −120°.

2. Find sin 780°.

3. Find sin 17π/3.

4. Find cos −15π/4.

5. Find sin 19π/6.

6. Find cos 99π/4.

7. Verify $\dfrac{1-2\cos^2\theta}{\sin\theta\cos\theta} = \tan\theta - \cot\theta$.

8. Verify $\dfrac{\sin\theta}{\sec\theta} = \dfrac{1}{\tan\theta + \cot\theta}$.

9. If the base of an isosceles triangle has length 10, express its area A as a function of the vertex angle θ.

10. Find the values of cos 11π/4 and sin 11π/4.

11. Find the three smallest positive numbers θ where sin $\theta = 1$.

12. Suppose $0 < \theta < \pi/2$ and cos $\theta = \frac{2}{5}$. Find sin θ.

13. Find the smallest positive number x such that $\sin(x^2 + x + 4) = 0$.

14. Suppose $0 < \theta < \pi/2$ and tan $\theta = 2$. Find cos θ.

15. Find the smallest number x such that tan $e^x = 0$.

4

Trig Values for Special Angles

Figure 4.1 shows the first-quadrant angles π/6 (30°), π/4 (45°), and π/3 (60°) in the unit circle, and values for sin θ, cos θ, and tan θ.

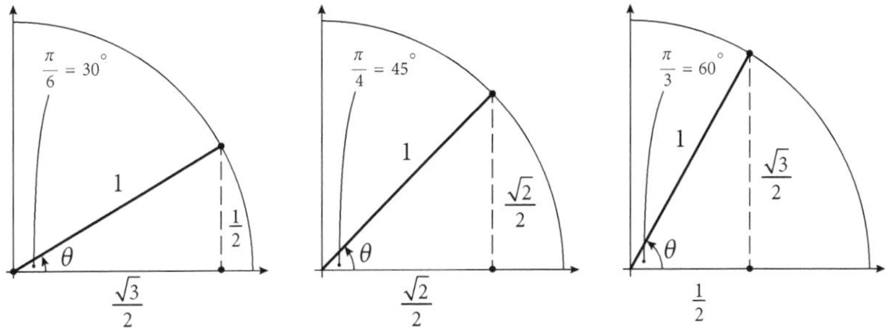

$$\sin\frac{\pi}{6} = \frac{1}{2} \qquad \cos\frac{\pi}{6} = \frac{\sqrt{3}}{2} \qquad \tan\frac{\pi}{6} = \frac{\frac{1}{2}}{\frac{\sqrt{3}}{2}} = \frac{1}{\sqrt{3}} = \frac{\sqrt{3}}{3}$$

$$\sin\frac{\pi}{4} = \frac{\sqrt{2}}{2} \qquad \cos\frac{\pi}{4} = \frac{\sqrt{2}}{2} \qquad \tan\frac{\pi}{4} = \frac{\frac{\sqrt{2}}{2}}{\frac{\sqrt{2}}{2}} = 1$$

$$\sin\frac{\pi}{3} = \frac{\sqrt{3}}{2} \qquad \cos\frac{\pi}{3} = \frac{1}{2} \qquad \tan\frac{\pi}{3} = \frac{\frac{\sqrt{3}}{2}}{\frac{1}{2}} = \sqrt{3}$$

Figure 4.1 Some special angles in the first quadrant

In Figure 4.1, the trig values for π/6, π/4, and π/3 are easy to find by using the definitions of sin θ, cos θ, and tan θ (page 20) and the facts in "Two Special Right Triangles" (page 4). By the definitions, we also know that sin 0 = 0, cos 0 = 1, and tan 0 = 0; and sin π/2 = 1, cos π/2 = 0, and tan π/2 is undefined (1/0). These values and a few others are combined in the following table.

θ	$\sin\theta$	$\cos\theta$	$\tan\theta$
0	0	1	0
$\dfrac{\pi}{6}$	$\dfrac{1}{2}$	$\dfrac{\sqrt{3}}{2}$	$\dfrac{\sqrt{3}}{3}$
$\dfrac{\pi}{4}$	$\dfrac{\sqrt{2}}{2}$	$\dfrac{\sqrt{2}}{2}$	1
$\dfrac{\pi}{3}$	$\dfrac{\sqrt{3}}{2}$	$\dfrac{1}{2}$	$\sqrt{3}$
$\dfrac{\pi}{2}$	1	0	undefined
$\dfrac{2\pi}{3}$	$\dfrac{\sqrt{3}}{2}$	$-\dfrac{1}{2}$	$-\sqrt{3}$
$\dfrac{3\pi}{4}$	$\dfrac{\sqrt{2}}{2}$	$-\dfrac{\sqrt{2}}{2}$	-1
$\dfrac{5\pi}{6}$	$\dfrac{1}{2}$	$-\dfrac{\sqrt{3}}{2}$	$-\dfrac{\sqrt{3}}{3}$
π	0	-1	0
$\dfrac{3\pi}{2}$	-1	0	undefined
2π	0	1	0

The values in the preceding table are best learned not by memorization but by drawing or visualizing the appropriate diagram and applying the definitions of the trig functions.

Signs of Trig Values by Quadrant

If you know the values of sin θ, cos θ, and tan θ for angles in the first quadrant, then you can find their values for angles in any quadrant. First, determine the sign of the function by the quadrant of θ. The following table summarizes signs by quadrant (see Figure 3.2 on page 20).

Quadrant	1	2	3	4
sin θ	+	+	–	–
cos θ	+	–	–	+
tan θ	+	–	+	–

Second, determine the positive acute angle θ' (the **reference angle**) formed by the terminal side of θ and the nearest half of the x-axis (Figure 4.2).

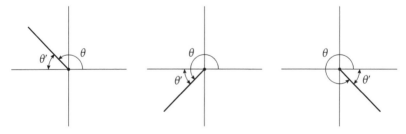

Figure 4.2 Determining the reference angle θ' from the angle θ

The desired value will be one of ±sin θ', ±cos θ', or ±tan θ', with the sign determined by the quadrant of θ. A few examples:

- If $\theta = 2\pi/3$ (120°), then $\theta' = \pi/3$ (60°). Because θ is in the second quadrant, cos $\theta = -\cos \theta' = -\cos \pi/3 = -\frac{1}{2}$.

- If $\theta = 5\pi/4$ (225°), then $\theta' = \pi/4$ (45°). Because θ is in the third quadrant, tan $\theta = +\tan \theta' = \tan \pi/4 = 1$.

- If $\theta = 11\pi/6$ (330°), then $\theta' = \pi/6$ (30°). Because θ is in the fourth quadrant, sin $\theta = -\sin \theta' = -\sin \pi/6 = -\frac{1}{2}$.

Problems

Find exact numerical values for the following expressions.

1. $\dfrac{\sin \dfrac{\pi}{2} + \cos \dfrac{\pi}{2}}{\sin \pi + \cos \pi}$

2. $\dfrac{1 + \tan^2 \dfrac{\pi}{3}}{1 + \cot^2 \dfrac{\pi}{3}}$

3. $\dfrac{\sin \pi + \cos(-\pi)}{\sin \dfrac{\pi}{2} + \cos\left(-\dfrac{\pi}{2}\right)}$

4. $\dfrac{\sin \pi \cos \pi \tan \pi}{\sin \dfrac{\pi}{3} \cos \dfrac{\pi}{3} \tan \dfrac{\pi}{3}}$

5. $\sin \dfrac{5\pi}{4} \sin \dfrac{3\pi}{4} \sin \dfrac{\pi}{4}$

5 Graphs of Trig Functions

By looking at Figure 5.1, you can draw sin θ by following how y varies as θ increases from 0 to 2π, that is, as the radius sweeps through a complete counterclockwise rotation. Figure 5.2 shows a complete cycle of sin θ.

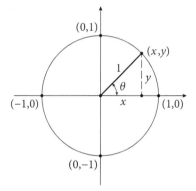

Figure 5.1 sin $\theta = y$, cos $\theta = x$, and tan $\theta = y/x$

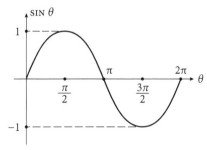

Figure 5.2 Complete cycle of sin θ

Figure 5.3 shows the complete graph of sin θ, which consists of infinitely many repetitions of the cycle in Figure 5.2, to the left and to the right.

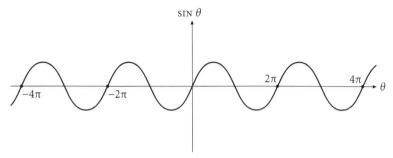

Figure 5.3 Graph of sin θ

The graph of cos θ is drawn essentially in the same way (Figure 5.4 and Figure 5.5). The main difference is that cos θ starts at 1 when $\theta = 0$.

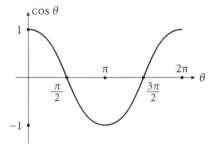

Figure 5.4 Complete cycle of cos θ

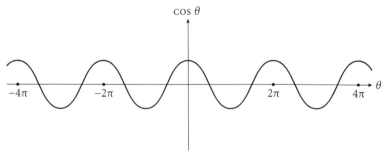

Figure 5.5 Graph of cos θ

Period

In Figure 5.1, the angles θ and $\theta + 2\pi$ are coterminal angles (page 16), meaning that they have the same terminal sides. For this reason, sin θ and cos θ are **periodic** with **period** 2π. Algebraically

$$\sin(\theta + 2\pi) = \sin(\theta)$$

and

$$\cos(\theta + 2\pi) = \cos(\theta)$$

If b is a positive constant, then the function sin $b\theta$ completes one cycle as $b\theta$ increases from 0 to 2π, that is, when θ increases from 0 to $2\pi/b$. This function is periodic with period $2\pi/b$.

Amplitude

To magnify the graph of sin $b\theta$ by a factor of a in the vertical direction, multiply by the positive constant a, called the **amplitude**, giving

$$a \sin b\theta$$

The peaks of a sin $b\theta$ are a units above the θ-axis.

Figure 5.6 shows the graph of a sin 2θ from 0 to 2π. This graph has an amplitude of a (the graph ranges between $-a$ and a). Its period is $2\pi/\pi = \pi$, so it makes one complete cycle in the interval from 0 to π. The graph of a sin 2θ oscillates twice as fast as a sin θ.

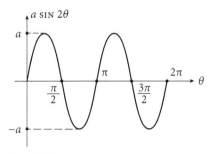

Figure 5.6 Graph of a sin 2θ

By comparison, the graph of a sin $\frac{1}{2}\theta$ oscillates *half* as fast as a sin θ, making one complete cycle in the interval from 0 to 4π. The graph of a sin $\frac{1}{2}\theta$ is a "stretched" version of the graph of a sin θ.

Graphing tan θ

Figure 5.7 is Figure 5.1 with a vertical line through the point (1, 0) added, tangent to the circle at that point.

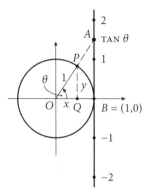

Figure 5.7 Geometric interpretation of $\tan \theta = y/x = PQ/OQ$

A given angle θ has a terminal side that doesn't coincide with the *y*-axis. Extend the terminal side of θ, either forward or backward through the origin *O*, until it intersects the auxiliary vertical line at point *A*. The *y*-coordinate of *A* is the value of tan θ. To see this fact, note that *OPQ* and *OAB* are similar triangles (page 3), so

$$\frac{PQ}{OQ} = \frac{AB}{OB} = \frac{AB}{1} = AB$$

As θ increases from −π/2 to π/2, the point *A* moves up the line from −∞ through 0 to +∞. The graph (Figure 5.8) is periodic with period π: $\tan(\theta + \pi) = \tan(\theta)$. The range excludes odd multiples of π/2.

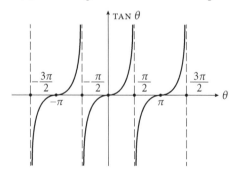

Figure 5.8 Graph of tan θ

Problems

1. What is range of $4 \sin x$ and of $\sin(4x)$?

2. What are the range, amplitude, and period of $\sin^2 x$?

6

The Major Formulas

This chapter completes the list of trigonometric identities started in Chapter 3 (see "The Essential Trigonometric Identities" on page 21). For the following identities, the Greek letters θ and φ (theta and phi) denote two arbitrary angles.

Addition Formulas

The **addition formulas** are

$$\sin(\theta + \varphi) = \sin\theta\cos\varphi + \cos\theta\sin\varphi \qquad (12)$$

$$\cos(\theta + \varphi) = \cos\theta\cos\varphi - \sin\theta\sin\varphi \qquad (13)$$

$$\tan(\theta + \varphi) = \frac{\tan\theta + \tan\varphi}{1 - \tan\theta\tan\varphi} \qquad (14)$$

To prove (14), express $\tan(\theta + \varphi)$ in terms of $\sin(\theta + \varphi)$ and $\cos(\theta + \varphi)$.

$$\tan(\theta + \varphi) = \frac{\sin(\theta + \varphi)}{\cos(\theta + \varphi)}$$

$$= \frac{\sin\theta\cos\varphi + \cos\theta\sin\varphi}{\cos\theta\cos\varphi - \sin\theta\sin\varphi} \qquad \text{divide both numerator and denominator by } \cos\theta\cos\varphi$$

$$= \frac{\dfrac{\sin\theta}{\cos\theta} + \dfrac{\sin\varphi}{\cos\varphi}}{1 - \left(\dfrac{\sin\theta}{\cos\theta}\right)\left(\dfrac{\sin\varphi}{\cos\varphi}\right)}$$

$$= \frac{\tan\theta + \tan\varphi}{1 - \tan\theta\tan\varphi}$$

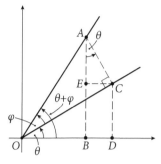

Figure 6.1 Geometric proof of (12) and (13)

To prove (12) and (13), refer to Figure 6.1, in which

- The angles θ and φ are positive acute angles whose sum is also acute
- The arbitrary point A lies on the terminal side of $\theta + \varphi$
- AB is perpendicular to the x-axis
- AC is perpendicular to the terminal side of θ
- CD is perpendicular to the x-axis
- CE is perpendicular to AB
- The angle EAC equals θ (because both are acute angles and their sides are respectively perpendicular)

The proof of (12) is

$$\sin(\theta + \varphi) = \frac{AB}{OA} = \frac{AE + EB}{OA} = \frac{AE + CD}{OA}$$

$$= \frac{AE}{OA} + \frac{CD}{OA}$$

$$= \frac{AE}{AC} \cdot \frac{AC}{OA} + \frac{CD}{OC} \cdot \frac{OC}{OA}$$

$$= \cos\theta \sin\varphi + \sin\theta \cos\varphi$$

The proof of (13) is

$$\cos(\theta + \varphi) = \frac{OB}{OA} = \frac{OD - BD}{OA} = \frac{OD - EC}{OA}$$

$$= \frac{OD}{OA} - \frac{EC}{OA}$$

$$= \frac{OD}{OC} \cdot \frac{OC}{OA} - \frac{EC}{AC} \cdot \frac{AC}{OA}$$

$$= \cos\theta \cos\varphi - \sin\theta \sin\varphi$$

Subtraction Formulas

The **subtraction formulas** are

$$\sin(\theta - \varphi) = \sin\theta \cos\varphi - \cos\theta \sin\varphi \qquad (15)$$

$$\cos(\theta - \varphi) = \cos\theta \cos\varphi + \sin\theta \sin\varphi \qquad (16)$$

$$\tan(\theta - \varphi) = \frac{\tan\theta - \tan\varphi}{1 + \tan\theta \tan\varphi} \qquad (17)$$

The subtraction formulas can be proved by replacing φ by $-\varphi$ in the addition formulas and using (6), (7), and (8) (see "Identities for $-\theta$" on page 22).

The proof of (15), for example, is

$$\sin(\theta - \varphi) = \sin(\theta + (-\varphi))$$

$$= \sin\theta \cos(-\varphi) + \cos\theta \sin(-\varphi)$$

$$= \sin\theta \cos\varphi - \cos\theta \sin\varphi$$

Double-Angle Formulas

The **double-angle formulas** are

$$\sin 2\theta = 2\sin\theta \cos\theta \qquad (18)$$

$$\cos 2\theta = \cos^2\theta - \sin^2\theta \qquad (19)$$

(18) and (19) can be proved as special cases of (12) and (13), respectively, by replacing φ with θ to get $\sin(\theta + \theta)$ and $\cos(\theta + \theta)$.

The double-angle formula for tangent isn't used much in calculus:

$$\tan 2\theta = \frac{2\tan\theta}{1-\tan^2\theta}$$

Half-Angle Formulas

The **half-angle formulas** are

$$2\sin^2\theta = 1 - \cos 2\theta \qquad (20)$$

$$2\cos^2\theta = 1 + \cos 2\theta \qquad (21)$$

(20) and (21) can be rewritten in "true" half-angle form as

$$\sin\frac{\theta}{2} = \pm\sqrt{\frac{1-\cos\theta}{2}} \quad \text{and} \quad \cos\frac{\theta}{2} = \pm\sqrt{\frac{1+\cos\theta}{2}}$$

but in calculus problems, (20) and (21) are the more common and tractable forms of these identities.

To prove (20), subtract (19) from (9):

$$\begin{aligned}
\cos^2\theta + \sin^2\theta &= 1 \\
\cos^2\theta - \sin^2\theta &= \cos 2\theta \\
\hline
2\sin^2\theta &= 1 - \cos 2\theta
\end{aligned}$$

To prove (21), add (19) to (9):

$$\begin{aligned}
\cos^2\theta + \sin^2\theta &= 1 \\
\cos^2\theta - \sin^2\theta &= \cos 2\theta \\
\hline
2\cos^2\theta &= 1 + \cos 2\theta
\end{aligned}$$

The half-angle formula for tangent isn't used much in calculus:

$$\tan\frac{\theta}{2} = \frac{\sin\theta}{1+\cos\theta} = \frac{1-\cos\theta}{\sin\theta}$$

Problems

1. Use the subtraction formula to find $\cos \pi/12$.

2. Use the half-angle formula to find $\sin \pi/12$.

3. Find a formula for $\sin(4\theta)$ in terms of $\cos \theta$ and $\sin \theta$.

4. Find a formula for $\sin(3\theta)$ in terms of $\sin \theta$.

5. Verify $\cot\theta = \dfrac{\sin 2\theta}{1 - \cos 2\theta}$.

6. Verify $\cos 2\theta = \dfrac{1 - \tan^2 \theta}{1 + \tan^2 \theta}$.

7. Verify $\tan\theta = \dfrac{\sin 2\theta}{1 + \cos 2\theta}$.

8. Verify $\tan^2\theta = \dfrac{1 - \cos 2\theta}{1 + \cos 2\theta}$.

9. Verify $1 - 4\sin^4 \theta = \cos 2\theta\left(1 + 2\sin^2 \theta\right)$.

7 Inverse Trig Functions

The **inverse trigonometric functions** are the inverse functions of the trigonometric functions with suitably restricted domains.

Notation

If you know that $\sin \pi/6 = \frac{1}{2}$, then you therefore know that the angle (in radians) whose sine is $\frac{1}{2}$ is $\pi/6$. The symbols that denote an angle whose sine is a given number x are

$$\sin^{-1} x \qquad \text{and} \qquad \arcsin x$$

These expressions are equivalent and fully interchangeable. The first is read "the inverse sine of x" and the second is read "the arc sine of x" but both mean "an angle whose sine is x." Don't be confused by the inconsistency in common notation: the -1 is *not* an exponent, so $\sin^{-1} x$ never means $1/\sin(x)$ or $(\sin x)^{-1}$.

The formulas

$$x = \sin y \qquad \text{and} \qquad y = \sin^{-1} x$$

mean the same thing, just like the formulas $x = 2y$ and $y = x/2$ mean the same thing. In each case, the first equation is solved for x and the second (same) equation is solved for y.

Graphing arcsin x

To graph $y = \sin^{-1} x$, take the graph of $y = \sin x$ (see Figure 5.3 on page 30) and flip it by swapping the x and y values (Figure 7.1). Note that y exists only when x is in the interval $-1 \leq x \leq 1$. For any x in this interval, however, there exist infinitely many corresponding values of y. Recall from "Functions" on page 6 that a function requires each x to have *exactly one* corresponding y. To qualify as a function, the values of $y = \sin^{-1} x$ are arbitrarily restricted to the interval $-\pi/2 \leq y \leq \pi/2$, indicated by the heavy part of the graph in Figure 7.1. This restriction guarantees a single-valued solution when solving the equation $\sin y = x$ for y.

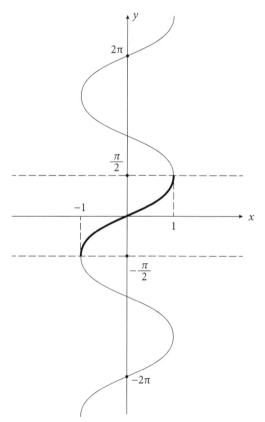

Figure 7.1 Graph of arcsin x (heavy part of the curve)

Graphing arctan *x*

The function $y = \tan^{-1} x$ (or $y = \arctan x$) is defined in about the same way:

$$y = \tan^{-1} x \qquad \text{means} \qquad \tan y = x \qquad \text{where} \qquad -\frac{\pi}{2} < y < \frac{\pi}{2}$$

The expression $\tan^{-1} x$ is read "the inverse tangent of *x*" and means "the angle (in the specified range) whose tangent is *x*." The heavy curve in Figure 7.2 is the graph of $y = \tan^{-1} x$.

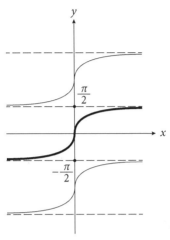

Figure 7.2 Graph of arctan *x* (heavy curve)

Other Inverse Trig Functions

The other four trig functions have inverses, but they're typically not needed in real problems. They're summarized in the following table.

Function	Domain	Range
$\sin^{-1}$	$-1 \leq x \leq 1$	$-\pi/2 \leq y \leq \pi/2$
$\cos^{-1}$	$-1 \leq x \leq 1$	$0 \leq y \leq \pi$
$\tan^{-1}$	all real numbers	$-\pi/2 < y < \pi/2$
$\cot^{-1}$	all real numbers	$0 < y < \pi$
$\sec^{-1}$	$x \leq -1$ or $1 \leq x$	$0 \leq y < \pi/2$ or $\pi/2 < y \leq \pi$
$\csc^{-1}$	$x \leq -1$ or $1 \leq x$	$-\pi/2 \leq y < 0$ or $0 < y \leq \pi/2$

Problems

1. Find a formula for $\tan(\cos^{-1}(x))$. Assume $-1 \le x \le 1$ with $x \ne 0$.

2. Evaluate $\cos^{-1}(\cos 3\pi)$.

3. Find exact numerical values for the following expressions.

$$\sin^{-1} 0 \qquad \sin^{-1} 1 \qquad \sin^{-1}\left(-\frac{\sqrt{3}}{2}\right)$$

$$\sin^{-1}\left(-\frac{1}{2}\right) \qquad \tan^{-1}\frac{\sqrt{3}}{3} \qquad \tan^{-1} 1$$

$$\tan^{-1}\left(-\sqrt{3}\right) \qquad \tan^{-1}\left(-\frac{\sqrt{3}}{3}\right) \qquad \cos\left(\sin^{-1} 1\right)$$

8 The Law of Cosines (and Sines)

The **law of cosines**, which plays a crucial role in physics and geometry, generalizes to all triangles the Pythagorean theorem (page 3), which holds for only right triangles.

The law of cosines expresses the third side c of a triangle in terms of two given sides a and b and the angle θ opposite side c:

$$c^2 = a^2 + b^2 - 2ab\cos\theta$$

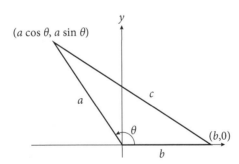

Figure 8.1 Law of cosines proof

To prove the law of cosines, place a triangle in the xy-plane as shown in Figure 8.1. Apply the distance formula (page 5) to the vertices $(b, 0)$ and $(a\cos\theta, a\sin\theta)$. The square of side c is

$$c^2 = (a\cos\theta - b)^2 + (a\sin\theta - 0)^2$$
$$= a^2(\cos^2\theta + \sin^2\theta) + b^2 - 2ab\cos\theta$$
$$= a^2 + b^2 - 2ab\cos\theta$$

The **law of sines** (the less-useful cousin of the law of cosines) states

$$\frac{\sin A}{a} = \frac{\sin B}{b} = \frac{\sin C}{c}$$

in a triangle with sides whose lengths are a, b, and c, with corresponding angles A, B, and C opposite those sides (Figure 8.2).

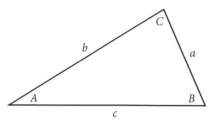

Figure 8.2 Reference triangle for the law of sines

To prove the law of sines, add the triangle's height h (Figure 8.3).

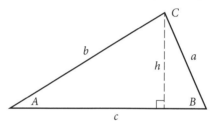

Figure 8.3 Law of sines proof

For the two resulting right triangles, we have

$$\sin A = \frac{h}{b} \qquad \text{and} \qquad \sin B = \frac{h}{a}$$

Rearranging terms yields $h = b \sin A$ and $h = a \sin B$. Equate these two expressions to h to get $b \sin A = a \sin B$ or equivalently

$$\frac{\sin A}{a} = \frac{\sin B}{b}$$

The remaining part of the law of sines is proved in the same way.

Note that $\sin A = h/b$. Hence $h = b \sin A$, so the area of triangle ABC is $\frac{1}{2}ch = \frac{1}{2}cb \sin A$. By similar derivations, the **area of a triangle** is

$$\frac{1}{2}bc \sin A = \frac{1}{2}ac \sin B = \frac{1}{2}ab \sin C$$

Problems

1. Suppose $a = 3$, $b = 5$, and $c = 6$. Find A, B, and C.

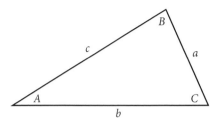

2. What is the largest possible area for a triangle that has one side of length 4 and one side of length 7?

9   Solutions

Chapter 1

1. The sum of the three angles in any triangle is 180°. In a right triangle, the right angle is 90°, so the two acute angles sum to 90°. Divide this amount into five equal parts of 18°. The smaller acute angle is 18° and the larger acute angle is $4 \times 18° = 72°$.

2. If the legs have length a, as shown in the following figure, then
$$h = \sqrt{a^2 + a^2} = \sqrt{2}a, \text{ so } a = \frac{h}{\sqrt{2}}. \text{ Hence the area is } A = \frac{1}{2}a^2 = \frac{1}{4}h^2.$$

3. As shown in the following figure, the height h satisfies
$$2^2 = h^2 + \left(\frac{x}{2}\right)^2$$

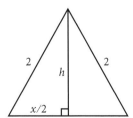

So

$$h = \sqrt{4 - \frac{x^2}{4}} = \frac{1}{2}\sqrt{16 - x^2}$$

Hence the area is

$$A = \frac{1}{2}xh = \frac{1}{2}x \cdot \frac{1}{2}\sqrt{16 - x^2} = \frac{1}{4}x\sqrt{16 - x^2}$$

4. Let y be the length of the other side, as shown in the following figure.

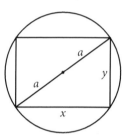

Therefore

$$x^2 + y^2 = (2a)^2$$

So

$$y = \sqrt{4a^2 - x^2}$$

Hence the area is

$$A = \frac{1}{2}xh = \frac{1}{2}x \cdot \frac{1}{2}\sqrt{16 - x^2} = \frac{1}{4}x\sqrt{16 - x^2}$$

5. The distance between $(-7, 3)$ and $(1, -2)$ is

$$\sqrt{(-7-1)^2 + (3-(-2))^2} = \sqrt{64 + 25} = \sqrt{89}$$

6. The distance between (a, b) and (b, a) is

$$\sqrt{(a-b)^2 + (b-a)^2} = \sqrt{2(a-b)^2} = \sqrt{2}|a-b|$$

7. If $f(x) = 5x^2 - 3$, then:

(a) $f(-3) = 5(-3)^2 - 3 = 42$

(b) $f(2) = 5 \times 2^2 - 3 = 17$

(c) $f(0) = 5 \times 0^2 - 3 = -3$

(d) $f(-\sqrt{7}) = 5f(-\sqrt{7})^2 - 3 = 32$

(e) $f(a + 3) = 5(a + 3)^2 - 3 = 5a^2 + 30a + 42$

(f) $f(5t) = 5(5t)^2 - 3 = 125t^2 - 3$

8. (a) $\dfrac{f(x+h)-f(x)}{h} = \dfrac{[5(x+h)-3]-[5x-3]}{h} = \dfrac{5h}{h} = 5$

(b) $\dfrac{f(x+h)-f(x)}{h} = \dfrac{(x+h)^2 - x^2}{h} = \dfrac{2xh+h^2}{h} = 2x+h$

(c) $\dfrac{f(x+h)-f(x)}{h} = \dfrac{\dfrac{1}{x+h} - \dfrac{1}{x}}{h} = \dfrac{x-(x+h)}{hx(x+h)} = -\dfrac{1}{x(x+h)}$

9. (a) No; $f(x) = x/x$ and $g(x) = 1$ are not equal because x/x is not defined for $x = 0$.

(b) Yes; $f(x) = x^2 - 1$ and $g(x) = (x + 1)(x - 1)$ are equal.

(c) No; $f(x) = x$ and $g(x) = \sqrt{x^2} = |x|$ don't agree for $x < 0$.

(d) No; $f(x) = x$ and $g(x) = (\sqrt{x})^2$ don't agree for $x < 0$, where $g(x)$ is not defined.

Chapter 2

1. To convert from degrees to radians, multiply by $\pi/180$.

$$15° = 15 \cdot \frac{\pi}{180} = \frac{\pi}{12} \qquad 150° = 150 \cdot \frac{\pi}{180} = \frac{5\pi}{6} \qquad 1500° = \frac{50\pi}{6}$$

$$-36° = -\frac{\pi}{5} \qquad\qquad -45° = -\frac{\pi}{4} \qquad\qquad 7° = \frac{7\pi}{180}$$

$$900° = 5\pi \qquad\qquad 1080° = 6\pi \qquad\qquad -110° = -\frac{11\pi}{18}$$

2. To convert from radians to degrees, multiply by $180/\pi$.

$$4\pi = 4\pi \cdot \frac{180}{\pi} = 720° \qquad \frac{\pi}{9} = \frac{\pi}{9} \cdot \frac{180}{\pi} = 20° \qquad 3 = 3 \cdot \frac{180}{\pi} = \frac{540°}{\pi}$$

$$-\frac{2\pi}{3} = -120° \qquad\qquad \frac{3\pi}{2} = 270° \qquad\qquad 2\pi = 360°$$

$$\frac{7\pi}{6} = 210° \qquad\qquad \frac{14\pi}{6} = 420° \qquad\qquad -5\pi = -900°$$

3. Pizzas are measured by their diameter. The radius of the pizza is 8 inches. The area of the slice is $\frac{1}{2}\theta r^2 = \frac{1}{2} \times \frac{3}{4} \times 8^2 = 24$ square inches.

4. The radius whose endpoint equals $(0, 1)$ makes an angle of $\pi/2$ radians ($90°$) with the positive x-axis. This angle is the smaller angle in the following figure.

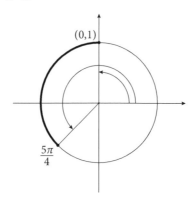

Because $5\pi/4 = \pi + \pi/4$, the radius that forms an angle of $5\pi/4$ radians with the positive x-axis lies $\pi/4$ radians beyond the negative x-axis (half-way between the negative x-axis and the negative y-axis). Hence the particle stops at the endpoint of the radius of the larger angle in the figure.

The particle travels along the heavy circular arc shown in the figure. This circular arc corresponds to an angle of $5\pi/4 - \pi/2 = 3\pi/4$ radians, so the particle travels $3\pi/4$ radians.

Chapter 3

1. Note that $-120° = -2\pi/3$ is one-third of the way from $-\pi/2$ to $-\pi$, so the endpoint on the unit circle is in the third quadrant with coordinates $(-\frac{1}{2}, -\frac{\sqrt{3}}{2})$, so $\cos -120° = -\frac{1}{2}$.

2. Because $\sin \theta$ is periodic with period $2\pi = 360°$,

$$\sin 780° = \sin(2 \cdot 360° + 60°) = \sin 60° = \sin \frac{\pi}{3} = \frac{\sqrt{3}}{2}$$

3. Because $\sin \theta$ is periodic with period 2π,

$$\sin \frac{17\pi}{3} = \sin\left(2 \cdot 2\pi + \frac{5\pi}{3}\right) = \sin \frac{5\pi}{3} = -\frac{\sqrt{3}}{2}$$

4. Because $\cos \theta$ is periodic with period 2π,

$$\cos \frac{-15\pi}{4} = \cos\left(\frac{\pi}{4} - 2 \cdot 2\pi\right) = \cos \frac{\pi}{4} = \frac{\sqrt{2}}{2}$$

5. Because $\sin \theta$ is periodic with period 2π,

$$\sin \frac{19\pi}{6} = \sin\left(2\pi + \frac{7\pi}{6}\right) = \sin \frac{7\pi}{6}$$

Note that $7\pi/6$ is one-third of the way from π to $3\pi/2$, so the endpoint on the unit circle is in the third quadrant with coordinates $(-\frac{\sqrt{3}}{2}, -\frac{1}{2})$, so $\sin 19\pi/6 = -\frac{1}{2}$.

6. Because $\cos\theta$ is periodic with period 2π,

$$\cos\frac{99\pi}{4} = \cos\left(12 \cdot 2\pi + \frac{3\pi}{4}\right) = \cos\frac{3\pi}{4} = -\frac{\sqrt{2}}{2}$$

7. $\tan\theta - \cot\theta = \dfrac{\sin\theta}{\cos\theta} - \dfrac{\cos\theta}{\sin\theta}$

$$= \frac{\sin^2\theta - \cos^2\theta}{\sin\theta\cos\theta}$$

$$= \frac{\left(1-\cos^2\theta\right) - \cos^2\theta}{\sin\theta\cos\theta}$$

$$= \frac{1 - 2\cos^2\theta}{\sin\theta\cos\theta}$$

8. $\dfrac{1}{\tan\theta + \cot\theta} = \dfrac{1}{\dfrac{\sin\theta}{\cos\theta} + \dfrac{\cos\theta}{\sin\theta}}$

$$= \frac{1}{\dfrac{\sin^2\theta + \cos^2\theta}{\sin\theta\cos\theta}}$$

$$= \frac{\sin\theta\cos\theta}{\sin^2\theta + \cos^2\theta}$$

$$= \sin\theta\cos\theta$$

$$= \frac{\sin\theta}{1/\cos\theta}$$

$$= \frac{\sin\theta}{\sec\theta}$$

9. If h is the height, then $h/5 = \cot \theta/2$, so

$$h = 5\cot\frac{\theta}{2} \text{ and } A = \frac{1}{2}bh = \frac{1}{2} \cdot 10 \cdot 5\cot\frac{\theta}{2} = 25\cot\frac{\theta}{2}$$

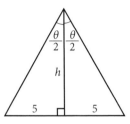

10. Because $11\pi/4 = 2\pi + \pi/2 + \pi/4$, an angle of $11\pi/4$ radians consists of a complete rotation around the circle (2π radians) followed by another $\pi/2$ radians (90°), followed by another $\pi/4$ radians (45°). Hence the endpoint of the corresponding radius is $(-\frac{\sqrt{2}}{2}, \frac{\sqrt{2}}{2})$, so $\cos 11\pi/4 = -\frac{\sqrt{2}}{2}$ and $\sin 11\pi/4 = \frac{\sqrt{2}}{2}$.

11. As the radius of the unit circle sweeps counterclockwise from the positive x-axis, it first forms angles where $\sin \theta = 1$ at $\pi/2$, $5\pi/2$, and $9\pi/2$.

12. We know that

$$\cos^2 \theta + \sin^2 \theta = 1$$

Therefore

$$\sin^2 \theta = 1 - \cos^2 \theta = 1 - \left(\frac{2}{5}\right)^2 = \frac{21}{25}$$

Because $0 < \theta < \pi/2$, we know that $\sin \theta > 0$. Taking square roots of both sides, we have

$$\sin\theta = \frac{\sqrt{21}}{5}$$

13. Note that $x^2 + x + 4$ is an increasing function on the interval $[0, \infty)$. If x is positive, then $x^2 + x + 4 > 4$. Because $\pi < 4 < 2\pi$, the smallest number larger than 4 whose sine equals 0 is 2π. Therefore we want to choose x so that $x^2 + x + 4 = 2\pi$, meaning we must solve the equation

$$x^2 + x + (4 - 2\pi) = 0$$

The quadratic formula yields the solutions

$$x = \frac{-1 \pm \sqrt{8\pi - 15}}{2}$$

Choosing the plus sign gives a positive x, whereas choosing the minus sign gives a negative x. The problem calls for a positive number, so choose the plus sign.

14. We know that

$$\cos^2 \theta + \sin^2 \theta = 1$$

and

$$2 = \tan \theta = \frac{\sin \theta}{\cos \theta}$$

To find $\cos \theta$, substitute

$$\sin \theta = \sqrt{1 - \cos^2 \theta}$$

in the preceding equation (this substitution is valid because $0 < \theta < \pi/2$ and so $\sin \theta > 0$) to get

$$2 = \frac{\sqrt{1 - \cos^2 \theta}}{\cos \theta}$$

Square both sides and then multiply both sides by $\cos^2 \theta$ to get

$$5 \cos^2 \theta = 1$$

Because $0 < \theta < \pi/2$, we know that $\cos \theta > 0$. Taking square roots of both sides yields

$$\cos \theta = \frac{1}{\sqrt{5}}$$

15. Note that e^x is an increasing function. Because e^x is positive for every real number x, and because π is the smallest positive number whose tangent equals 0, choose x so that $e^x = \pi$. Therefore, $x = \log \pi$ (the natural log of π).

Chapter 4

1. $\dfrac{\sin \dfrac{\pi}{2} + \cos \dfrac{\pi}{2}}{\sin \pi + \cos \pi} = \dfrac{1+0}{0+-1} = -1$

2. $\dfrac{1 + \tan^2 \dfrac{\pi}{3}}{1 + \cot^2 \dfrac{\pi}{3}} = \dfrac{1+3}{1+\frac{1}{3}} = 3$

3. $\dfrac{\sin \pi + \cos(-\pi)}{\sin \dfrac{\pi}{2} + \cos\left(-\dfrac{\pi}{2}\right)} = \dfrac{0+-1}{1+0} = -1$

4. $\dfrac{\sin \pi \cos \pi \tan \pi}{\sin \dfrac{\pi}{3} \cos \dfrac{\pi}{3} \tan \dfrac{\pi}{3}} = \dfrac{0 \cdot -1 \cdot 0}{\dfrac{\sqrt{3}}{2} \cdot \dfrac{1}{2} \cdot \sqrt{3}} = 0$

5. $\sin \dfrac{5\pi}{4} \sin \dfrac{3\pi}{4} \sin \dfrac{\pi}{4} = -\dfrac{1}{\sqrt{2}} \cdot \dfrac{1}{\sqrt{2}} \cdot \dfrac{1}{\sqrt{2}} = -\dfrac{1}{2\sqrt{2}} = -\dfrac{\sqrt{2}}{4}$

Chapter 5

1. The range of 4 sin x is $[-4, 4]$ (multiply each number in the range of sin x by 4).

    The range of $\sin(4x)$ is $[-1, 1]$ (the same range as sin x).

2. The sine function takes on all values in the interval $[-1, 1]$. Squaring the numbers in this interval gives the numbers in the interval $[0, 1]$. The range of $\sin^2 x$ is $[0, 1]$.

    The function $\sin^2 x$ has a maximum value of 1 and a minimum value of 0. The difference between these extreme values is 1. Hence the amplitude of $\sin^2 x$ is ½.

    We know that $\sin(x + \pi) = -\sin x$. Squaring both sides of this equation, we get $\sin^2(x + \pi) = \sin^2 x$. No positive number p smaller than π can produce the identity $\sin^2(x + p) = \sin^2 x$, which can be seen by setting $x = 0$ to make the equation $\sin^2(p) = 0$. The smallest positive number p satisfying this equation is π. Hence the period of $\sin^2 x$ is π.

Chapter 6

1. $\cos\left(\dfrac{\pi}{12}\right) = \cos\left(\dfrac{\pi}{3} - \dfrac{\pi}{4}\right)$

    $= \cos\dfrac{\pi}{3}\cos\dfrac{\pi}{4} + \sin\dfrac{\pi}{3}\sin\dfrac{\pi}{4}$

    $= \dfrac{1}{2}\cdot\dfrac{\sqrt{2}}{2} + \dfrac{\sqrt{3}}{2}\cdot\dfrac{\sqrt{2}}{2}$

    $= \dfrac{\sqrt{2} + \sqrt{6}}{4}$

2. Use the half-angle formula for sin $\theta/2$ with $\theta = \pi/6$. Choose the positive square root because sin $\pi/12$ is positive.

    $$\sin\frac{\pi}{12} = \sqrt{\frac{1 - \cos\dfrac{\pi}{6}}{2}} = \sqrt{\frac{1 - \dfrac{\sqrt{3}}{2}}{2}} = \sqrt{\frac{\left(1 - \dfrac{\sqrt{3}}{2}\right)\cdot 2}{2\cdot 2}} = \frac{\sqrt{2 - \sqrt{3}}}{2}$$

3. Use the double-angle formula for sine, with θ replaced by 2θ, to get $\sin(4\theta) = 2\cos(2\theta)\sin(2\theta)$. Applying the double-angle formula for the right-side expressions yields

$$\sin(4\theta) = 2(2\cos^2\theta - 1)(2\cos\theta\sin\theta)$$
$$= 4(2\cos^2\theta - 1)\cos\theta\sin\theta$$

4. $\sin 3\theta = \sin(2\theta + \theta)$

$$= \sin 2\theta \cos\theta + \cos 2\theta \sin\theta$$

$$= (2\sin\theta\cos\theta)\cos\theta + (\cos^2\theta - \sin^2\theta)\sin\theta$$

$$= 2\sin\theta\cos^2\theta + \sin\theta\cos^2\theta - \sin^3\theta$$

$$= 3\sin\theta\cos^2\theta - \sin^3\theta$$

$$= 3\sin\theta(1 - \sin^2\theta) - \sin^3\theta$$

$$= 3\sin\theta - 4\sin^3\theta$$

5. $\dfrac{\sin 2\theta}{1 - \cos 2\theta} = \dfrac{2\sin\theta\cos\theta}{2\sin^2\theta} = \dfrac{\cos\theta}{\sin\theta} = \cot\theta$

6. $\cos 2\theta = \cos^2\theta - \sin^2\theta$

$$= \cos^2\theta\left(1 - \frac{\sin^2\theta}{\cos^2\theta}\right)$$

$$= \cos^2\theta(1 - \tan^2\theta)$$

$$= \frac{1 - \tan^2\theta}{\sec^2\theta}$$

$$= \frac{1 - \tan^2\theta}{1 + \tan^2\theta}$$

7. $\tan\theta = \dfrac{\sin\theta}{\cos\theta}$

   $= \dfrac{\sin\theta}{\cos\theta} \cdot \dfrac{2\cos\theta}{2\cos\theta}$

   $= \dfrac{\sin 2\theta}{2\cos^2\theta}$ double-angle formula used

   $= \dfrac{\sin 2\theta}{1+\cos 2\theta}$ half-angle formula used

8. $\tan^2\theta = \dfrac{\sin^2\theta}{\cos^2\theta}$

   $= \dfrac{\frac{1}{2}(1-\cos 2\theta)}{\frac{1}{2}(1+\cos 2\theta)}$

   $= \dfrac{1-\cos 2\theta}{1+\cos 2\theta}$ half-angle formula used

9. $1-4\sin^4\theta = 1-\left(2\sin^2\theta\right)^2$

   $= \left(1-2\sin^2\theta\right)\left(1+2\sin^2\theta\right)$

   $= \cos 2\theta\left(1+2\sin^2\theta\right)$ half-angle formula used

Chapter 7

1. Let $\theta = \cos^{-1} x$. Hence θ is in $[0, \pi]$ and $\cos \theta = x$. Now

$$\tan(\cos^{-1} x) = \tan \theta = \frac{\sin \theta}{\cos \theta} = \frac{\sqrt{1 - \cos^2 \theta}}{\cos \theta} = \frac{\sqrt{1 - x^2}}{x}$$

2. Because $\cos 3\pi = -1$, we know that

$$\cos^{-1}(\cos 3\pi) = \cos^{-1}(-1)$$

Because $\cos \pi = -1$, we know that $\cos^{-1}(-1) = \pi$. ($\cos 3\pi$ also equals -1, but $\cos^{-1}(-1)$ must be in the interval $[0, \pi]$). Hence $\cos^{-1}(\cos 3\pi) = \pi$.

3. $\sin^{-1} 0 = 0$ $\qquad \sin^{-1} 1 = \dfrac{\pi}{2}$ $\qquad \sin^{-1}\left(-\dfrac{\sqrt{3}}{2}\right) = -\dfrac{\pi}{3}$

$\sin^{-1}\left(-\dfrac{1}{2}\right) = -\dfrac{\pi}{6}$ $\qquad \tan^{-1}\dfrac{\sqrt{3}}{3} = \dfrac{\pi}{6}$ $\qquad \tan^{-1} 1 = \dfrac{\pi}{4}$

$\tan^{-1}\left(-\sqrt{3}\right) = -\dfrac{\pi}{3}$ $\qquad \tan^{-1}\left(-\dfrac{\sqrt{3}}{3}\right) = -\dfrac{\pi}{6}$ $\qquad \cos(\sin^{-1} 1) = 0$

Chapter 8

1. To find A, use the law of cosines in the form $a^2 = b^2 + c^2 - 2bc \cos A$.

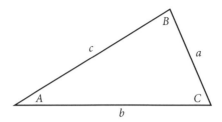

So $3^2 = 5^2 + 6^2 - 2 \cdot 5 \cdot 6 \cos A$. Solve for $\cos A$ to get $\cos A = {}^{13}\!/_{15}$.
Therefore $A = \cos^{-1}({}^{13}\!/_{15}) \approx 0.522$ radians $\approx 29.9°$.

To find B, use $b^2 = a^2 + c^2 - 2ac \cos B$.

To find C, use $c^2 = a^2 + b^2 - 2ab \cos C$.

2. Let θ be the angle between the two sides of lengths 4 and 7. Therefore the area of the triangle is $14 \sin \theta$ (by the law of sines). To choose θ to make the area as large as possible, note that the largest possible value of $\sin \theta$ is 1, which occurs when $\theta = \pi/2$. Hence we choose $\theta = \pi/2$, which gives us a right triangle with sides of lengths 4 and 7 around the right angle.

10

Trig Cheat Sheet

Radians-Degrees Conversions

$$\theta \text{ radians} = \left(\frac{180\theta}{\pi}\right)^{\circ} \quad \text{and} \quad \theta^{\circ} = \frac{\theta\pi}{180} \text{ radians}$$

Definitions of Trigonometric Functions

$\sin\theta = y$

$\cos\theta = x$

$\tan\theta = \dfrac{y}{x}$

$\cot\theta = \dfrac{x}{y}$

$\sec\theta = \dfrac{1}{x}$

$\csc\theta = \dfrac{1}{y}$

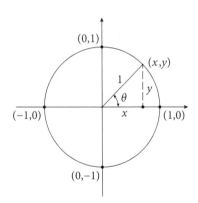

Basic Trigonometric Identities

$$\tan\theta = \frac{\sin\theta}{\cos\theta} \tag{1}$$

$$\cot\theta = \frac{\cos\theta}{\sin\theta} \tag{2}$$

$$\sec\theta = \frac{1}{\cos\theta} \tag{3}$$

$$\csc\theta = \frac{1}{\sin\theta} \tag{4}$$

$$\cot\theta = \frac{1}{\tan\theta} \tag{5}$$

$$\sin(-\theta) = -\sin(\theta) \tag{6}$$

$$\cos(-\theta) = \cos(\theta) \tag{7}$$

$$\tan(-\theta) = -\tan(\theta) \tag{8}$$

$$\sin^2\theta + \cos^2\theta = 1 \tag{9}$$

$$\tan^2\theta + 1 = \sec^2\theta \tag{10}$$

$$1 + \cot^2\theta = \csc^2\theta \tag{11}$$

Addition Formulas

$$\sin(\theta + \varphi) = \sin\theta\cos\varphi + \cos\theta\sin\varphi \qquad (12)$$

$$\cos(\theta + \varphi) = \cos\theta\cos\varphi - \sin\theta\sin\varphi \qquad (13)$$

$$\tan(\theta + \varphi) = \frac{\tan\theta + \tan\varphi}{1 - \tan\theta\tan\varphi} \qquad (14)$$

Subtraction Formulas

$$\sin(\theta - \varphi) = \sin\theta\cos\varphi - \cos\theta\sin\varphi \qquad (15)$$

$$\cos(\theta - \varphi) = \cos\theta\cos\varphi + \sin\theta\sin\varphi \qquad (16)$$

$$\tan(\theta - \varphi) = \frac{\tan\theta - \tan\varphi}{1 + \tan\theta\tan\varphi} \qquad (17)$$

Double-Angle Formulas

$$\sin 2\theta = 2\sin\theta\cos\theta \qquad (18)$$

$$\cos 2\theta = \cos^2\theta - \sin^2\theta \qquad (19)$$

Half-Angle Formulas

$$2\sin^2\theta = 1 - \cos 2\theta \qquad (20)$$

$$2\cos^2\theta = 1 + \cos 2\theta \qquad (21)$$

Law of Cosines and Law of Sines

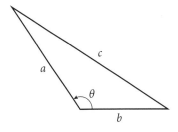 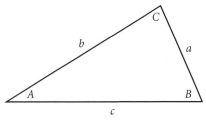

$$c^2 = a^2 + b^2 - 2ab\cos\theta$$

$$\frac{\sin A}{a} = \frac{\sin B}{b} = \frac{\sin C}{c}$$

Index

Made in the USA
Middletown, DE
05 July 2018